About This Report

Semiconductors have become an integral part of nearly every industry in advanced economies. The production of these semiconductors is largely centered in the western Pacific region and, for the highest-end semiconductors, exists almost entirely in Taiwan. Although this level of industrial base concentration for such a critical economic input raises challenges in and of itself, in this report, we examine the important geopolitical considerations of this concentration. The challenges around Taiwan's relationship with the People's Republic of China have not abated and have even accelerated to a degree. Taiwan's market dominance is enmeshed with these geopolitical considerations and requires careful analysis. To explore the geopolitical implications of Taiwan's semiconductor dominance, the RAND National Security Supply Chain Institute conducted a tabletop exercise with representatives from across the executive and legislative branches of the U.S. government and from a variety of industries that rely on semiconductors. Although this exercise did not produce definitive results, it did suggest several important—and, to a degree, counterintuitive—findings for continued exploration.

The research reported here was completed in August 2022 and underwent security review with the Defense Office of Prepublication and Security Review before public release.

RAND National Security Research Division

This research was conducted within the Navy and Marine Forces Program of the RAND National Security Research Division (NSRD), which operates the RAND National Defense Research Institute (NDRI), a federally funded research and development center (FFRDC) sponsored by the Office of the Secretary of Defense, the Joint Staff, the Unified Combatant Commands, the Navy, the Marine Corps, the defense agencies, and the defense intelligence enterprise.

For more information on the RAND Navy and Marine Forces Program, see www.rand.org/nsrd/nmf or contact the director (contact information is provided on the webpage).

Funding

Funding for this research was made possible by the independent research and development provisions of RAND's contracts for the operation of its U.S. Department of Defense federally funded research and development centers.

Acknowledgments

We would like to thank the many participants who contributed their time and expertise to the tabletop exercise execution. We also want to recognize the help of Lisa Jaycox, who facilitated the funding of this research. We thank our reviewers, Eric Heginbotham and Steve Worman. We also thank Barbara Bicksler, who provided indispensable assistance in organizing and revising this report.

Summary

Semiconductors have become an integral part of nearly every industry in advanced economies. The production of these semiconductors is largely centered in the western Pacific region and, for the highest-end semiconductors, exists almost entirely in Taiwan. The Taiwan Semiconductor Manufacturing Company (TSMC) is the dominant manufacturer of semiconductors, producing 92 percent of all logic chips that are ten nanometers or smaller.[1] Although this level of industrial base concentration for such a critical economic input raises challenges in and of itself, in this report, we examine the important geopolitical considerations of this concentration. The challenges around Taiwan's relationship with the People's Republic of China (PRC) have not abated and have even accelerated to a degree. Taiwan's market dominance is enmeshed with these geopolitical considerations and requires careful analysis.

To explore the geopolitical implications of Taiwan's semiconductor dominance, the RAND National Security Supply Chain Institute conducted a tabletop exercise (TTX) with representatives from across the executive and legislative branches of the U.S. government and from a variety of industries that rely on semiconductors.

TSMC's Dominance Creates Vulnerability

TSMC's dominance over the advanced semiconductor market results both from some unique market conditions and the company's diligence and careful management. TSMC is a technically proficient company operating in a portion of the microelectronics supply chain that is very capital-intensive and thus unattractive to companies seeking an immediately high rate of return. TSMC also has received direct support from the government of Taiwan, which has put the company at the center of a supply chain that is

---

[1] Antonio Varas, Raj Varadarajan, Jimmy Goodrich, and Falan Yinug, *Strengthening the Global Semiconductor Supply Chain in an Uncertain Era*, Boston Consulting Group and Semiconductor Industry Association, April 2021.

vital to the world. Finally, TSMC has pursued a global foundry model with multiple customers, as opposed to the vertically integrated model pursued by Intel. TSMC's dominance is in many ways the natural culmination of market impulses.

The implications of such positioning are particularly significant when considering the continuing geopolitical tension over Taiwan's future as an autonomously governed state, even as a "one China" construct continues to guide the national policies of both the United States and the PRC. The United States has reaffirmed that even though there is "one China," it will not accept unification by force. Although "strategic ambiguity" is its official position,[2] the United States has plans to defend Taiwan militarily, and the protection of Taiwan's autonomy remains a goal.

The TTX Demonstrated That There Are Generally Only Bad Options for Responding to the PRC Attempting to Coerce Taiwan, Using Semiconductor Access as Leverage

The TTX involved two scenarios, both of which began with a common set of conditions in which the PRC, for geopolitical reasons, imposed a coercive quarantine on Taiwan, as outlined in a 2022 RAND report.[3] The scenarios diverged in Taiwan's response to the quarantine.

In the first scenario, U.S. industry players sought to continue business as usual while legislative and executive participants sought paths to alternative supply. In the second scenario, with significant disruption to the supply of semiconductors generally and almost complete elimination of the supply of the highest-end semiconductors, the economic disruption was dire and nearly immediate. The choices available to the United States and its allies in

---

[2] "U.S. Maintains 'Strategic Ambiguity' over Taiwan: Security Adviser," *Nikkei Asia*, July 23, 2022.

[3] Bradley Martin, Kristen Gunness, Paul DeLuca, and Melissa Shostak, *Implications of a Coercive Quarantine of Taiwan by the People's Republic of China*, RAND Corporation, RR-A1279-1, 2022.

any circumstance in which the PRC might deny access to Taiwanese semiconductors very rapidly devolved into the following options:

- accept the PRC's demands and effectively abandon the autonomy of Taiwan
- try to build alternatives to reliance on Taiwanese semiconductors, requiring many years of diminished economic output, possibly to the point of economic depression
- go to war to protect Taiwan and coerce the PRC to stop its quest for unification, which would also be expected to lead to global economic disruption.

In the TTX, the scenarios purposely excluded the option of war, but the remaining choices are hardly desirable.

Most Players Did Not Immediately Understand Either the Direness of the Possible Situation or the Challenges Associated with Reacting to It

The supply chain as used in the TTX does exist, but players in the TTX were surprised at the degree of interdependence, with the first reactions from government players being a demand for additional analysis on the supply chain under threat.

In the uncontested scenario, the government teams perceived a major threat to U.S. security, warranting action to protect intellectual property (IP) and otherwise restrict exposure to potential Chinese influence. Industry players, however, generally viewed the threat to business being more from the U.S. government taking ill-considered action to restrict access to semiconductor manufacturing in a reunified Taiwan rather than from the condition of TSMC's ownership (and thus a huge portion of global supply of semiconductor manufacturing) passing into the Chinese Communist Party's (CCP's) hands. In the contested scenario, the degree of interdependence was surprising to all groups, and the reactions from all groups tended toward a desire for more-extensive government involvement to prioritize diminished semiconductor supply: effectively, the government taking over the allocation of limited semiconductor fabrication resources in the United States until alternatives could be found.

As a general matter, the players were not aware of the time and expense that would be required to reorient an industry built up over decades. Some solutions—such as stockpiling or recycling—were offered but were not likely to be effective or even realistic. The capital expense of building new fabrication facilities (commonly referred to as *fabs*) was not appreciated, nor was the challenge of generating a new labor supply sufficient to operate these complex facilities.

The Economic Impact of a Severe Supply Chain Disruption Would Create a National Security Challenge

Semiconductors are present in effectively every sector of the U.S. economy and in every other advanced economy. The existing supply chain evolved to promote efficient production and distribution with minimal duplication, and it accordingly has put a focus on locating semiconductor components where they are most readily produced. For the most part, the interconnections work well, with steady improvement in technology and effective delivery to consumers. The existing situation reflects years of private-sector decisionmaking focusing on market forces and shareholder value.

Interconnected economies do, however, create vulnerability if such factors as geopolitics result in a disruption to the supply chain. If semiconductors are denied to technology industries, these industries can no longer count on continued technological improvement as a means to maintain growth and market share. If semiconductors are unavailable to other industries, production suffers and shortages develop, as seen in the commodity chips shortage experienced in the auto industry in 2021.[4]

This disruption would affect both the PRC and Western economies. Even if the PRC managed to completely secure the supply of chips, economic disruption in the rest of the world would lower demand for Chinese goods,

---

[4] Michael Wayland, "Chip Shortage Expected to Cost Auto Industry $210 Billion in Revenue in 2021," CNBC, September 23, 2021.

meaning that fewer consumers globally would have the means to consume goods produced by PRC companies. Every major economy would suffer, leading to the following questions:

1. Who would suffer more and most immediately?
2. Who would best be able to adjust and overcome the disruption?

Such adjustment would require time, capital, an available workforce, and, possibly, improvements in technology. We know that it would take two to five years for the United States and its allies to build and outfit sufficient fabrication capacity to offset the loss of Taiwan's production. And this timeline is based on optimistic assumptions about tooling, permitting, and the labor market. In contrast, China—an autocratic society better able to harness the whole of government and the economy to pursue objectives—could probably build infrastructure faster, but we have not done the analysis to assess how quickly China could replicate Western tooling or develop the required labor market. The issue may be less about how quickly the PRC could generate new capacity than how long it could stand the overall decline in global economic activity.

Economic Vulnerability Could Provide the PRC with an Asymmetric Advantage

Economic considerations among even the closest allies can lead to a reshaping of alliances if nation-state economies are at risk. In the scenarios described in this report, advantage derives from being able to cope with disruptions to semiconductors produced in and exported from Taiwan.

Peaceful Unification Scenario

In this scenario, China was able to acquire a significant portion of the semiconductor global capacity without major cost to itself, and the United States and its allies were faced with a near-term choice of accepting the dominance and continuing to work with Taiwanese companies now owned by China or imposing sanctions and trying to cope with the loss of high-end production capability.

Industry and government players brought different perspectives. Although the industry players were aware that the rules for dealing with an authoritarian regime would be different and potentially more restrictive than engaging with democratic Taiwan, they felt that there was little option but to continue relationships with the companies now under CCP control. The government players, conversely, felt that the risks associated with trading directly with CCP-dominated companies were enough to warrant strong and immediate action. However, in the absence of prior investment to create alternative fabrication capacity, none of the actions recommended would have reduced vulnerability, at least not in the near term.

The de facto result would be that the United States and its allies would have to accept the changed relationship and probably could not significantly reduce vulnerability for several years. Meanwhile, the PRC would stand to gain global influence as a result of possessing a near monopoly on at least the fabrication of the world's most sophisticated semiconductors. This expanded influence could be expected to fundamentally change the global balance of power.

Contested Unification Scenario

In the contested scenario, the situation very rapidly became dire as access to Taiwan's semiconductor manufacturing quickly disappeared. The United States and its allies had no good nonmilitary options for dealing with the disruption, and conflict resolution came down to who would be better able to absorb the economic impact. The industry and government perspectives converged: Industry went as far to say that government action would be needed to adjudicate prioritization of limited supply.

The choices rapidly became

- cease supporting the Taiwanese effort to resist the coercive quarantine so that capacity could be restored through a Chinese-controlled Taiwan
- support the Taiwanese resistance efforts and accept the loss of access to semiconductors and the loss of trade with the PRC, and thus face an economic depression
- consider military action to directly challenge and coerce the PRC.

Since the TTX excluded a military option, the choices in the contested scenario came down to capitulating and ceding significant influence to the PRC—and overriding the wishes of the Taiwanese people—or supporting Taiwan at the cost of an economic depression for most of the world.

Recommendations

Using the results of the TTX, we developed the following recommendations for the executive branch of the U.S. government, the U.S. Congress, the governments of U.S. allies and partners, and industries with equities in the semiconductor supply chain:

1. **Improve analysis and understanding of the semiconductor supply chain specifically and the overall level of supply chain interdependence in general.** Two related but different lines of effort are needed.
    a. First, the semiconductor supply chain is one in a list of supply chains whose poorly understood interdependencies were brought into focus by the coronavirus disease 2019 (COVID-19) pandemic. Similar conditions could be present in multiple other sectors, but the research to establish these interdependencies has not been done.
    b. Second, from a geopolitical perspective, planning scenarios involving conflict over Taiwan's autonomous status do not include the loss of Taiwanese semiconductor capacity as a likely consequence. This consequence deserves significant consideration.
2. **An immediate and concerted effort should be made to reduce the concentration of semiconductor production in Taiwan.** This condition is not only dangerous to the world's economic well-being, it also increases Taiwan's vulnerability.
    a. TSMC should be incentivized to distribute production out of Taiwan. This does not imply moving all production, nor does it necessarily imply transfer of ownership. It means relocating production to places with less geopolitical significance than Taiwan. Reducing the risk of semiconductor disruption because

of Chinese aggression would increase the willingness of the United States and its allies to support Taiwan should aggression occur. This should be a powerful incentive for Taiwan.

b. Irrespective of TSMC actions, the U.S. and allied governments should take action to strengthen semiconductor production. Action does not imply top-down direction for investment, at least not in every case, but it does involve creating incentives for investment and creating opportunities for workforce training and liberalized immigration. It probably also involves management of IP-sharing with a clearer eye toward the security impacts of sharing designs, even those without an obvious defense tie. There might be designs that should be accessible only to producers inside the United States or those of preferred allies.

3. **Movement of facilities and equipment to the PRC should be specifically discouraged and heavily regulated.** If markets are incentivized to invest in the PRC and sell advanced equipment to Chinese companies, both are likely to occur. Eliminating such incentives is likely to require coordination with allies and goes against the normal imperatives of a market economy. Incentives need to be structured in ways that industry will see as effective.

4. **Collaborative relationships with allied governments and industries are essential, even if these appear counter to the normal impulse to keep sectors separate.** The interdependencies created by supply chains are complicated and extensive; individual and collective interests intertwine to a degree that means neither market nor normal government decisionmaking will be sufficient. The relationship between public and private sectors will require careful management, as will relationships with allies who have their own public-private challenges. But the TTX reinforced that neat separations between public and private interests are simply not possible in this context.

Contents

Figures and Tables

Figures

Tables

CHAPTER 1

The Economic and Geopolitical Complications of Interdependence

Claims that economic interdependence has in some way changed international norms and behavior are not new. Indeed, prior to World War I, observers famously predicted that the cost of war had gotten so great that major power conflict would be unthinkable.[1] As it turned out, interdependence did not deter war over the next generation. The economic component of international competition is also well understood: Competition between economic systems is one of the underpinnings of Marxism, underscored in the Cold War by Soviet promises to "bury" the United States under the rapid expansion of Eastern bloc economies.[2] However, this economic competition involved little interdependence between the blocs: The U.S.-dominated international system of the Cold War proved more successful than the Soviet system.

Nevertheless, the prognostications about the cost of war might have been less wrong than premature. The world has become interdependent to an unprecedented degree. We will show this in more detail in this report as we describe the specifics of the semiconductor market. Although semiconductors are not the only commodity or product with a complicated impact on worldwide supply chains, the semiconductor example is, in some important ways, both unique and representative.

---

[1] Norman Angell, *The Great Illusion; A Study of the Relation of Military Power in Nations to their Economic and Social Advantage*, G.P. Putnam's Sons, 1910.

[2] Alexandra Guzeva, "'We Will Bury You': What Nikita Khrushchev Actually Meant," *Russia Beyond*, January 13, 2022.

Geopolitics and Economic Interdependence

Economic interdependence develops for many reasons, one of which is the incentive for private enterprise to find the most efficient way to create products. Companies perceive that some part of their product is best created in a location where labor and resource access is favorable; these companies master the challenge of transportation between different production locations and ultimately create a supply chain that optimizes product inputs. Such behavior not only maximizes the company's profits but also reduces the cost of end items to consumers and distributes capital to places that might otherwise remain impoverished. In addition, maintaining parts of the supply chain in a variety of overseas locations can ensure access to markets, which can be beneficial for companies and countries alike. A blanket condemnation of interdependence would not be well-founded.

However, interdependence does create vulnerabilities that need to be understood; in the case of semiconductors, the steady migration of advanced capability to Taiwan was both a market and political decision that has generated both leverage and vulnerability. The near-monopoly that Taiwanese companies have over parts of the semiconductor industry means that a shift in geopolitical considerations would result in both Taiwan and the rest of the world experiencing significant vulnerability.

The impact of disruptions in the semiconductor market are likely to be broad within the United States and worldwide. Many analyses exist and much discussion has been generated about the concentration of semiconductor fabrication capacity in Taiwan and the lack of capacity (particularly for high-end semiconductors) in the United States.[3] This concentration has implications for U.S. strategic competition with China. However, the importance of semiconductors in the broader economy means that strategic competition should be framed more broadly than its potential effect on military or political outcomes. Actors in every sector across the United States are stakeholders. In this report, we explore the vulnerabilities of the semi-

---

[3] For example, see Antonio Varas, Raj Varadarajan, Jimmy Goodrich, and Falan Yinug, *Strengthening the Global Semiconductor Supply Chain in an Uncertain Era*, Boston Consulting Group and Semiconductor Industry Association, April 2021.

conductor supply chain and the impact of supply chain disruption in more detail through the results of a tabletop exercise (TTX).

Examining the Impact of Semiconductor Supply Chain Disruption Requires New Approaches and Perspectives

Although the official U.S. position toward Taiwan is strategic ambiguity, top U.S. leaders have stated that the United States would militarily defend Taiwan.[4] China's desired unification with Taiwan could lead to disruptions in the fabrication of semiconductors in Taiwan, which would have ripple effects across the U.S. and global economies. To better understand the implications of such disruptions and identify planning steps needed to mitigate risks, we analyzed the semiconductor supply chain (drawing on the work of many others in this area) and conducted a TTX that examined two scenarios of unification from broad national perspectives: private industry and the executive and legislative branches of government. By involving a wide variety of stakeholders, we were able to better understand the implications of potential disruptions for the United States and its allies from a holistic perspective.

Although many scholars and institutions have explored the potential Chinese annexation of Taiwan from a military perspective, far fewer have examined this scenario through an economic or diplomatic lens.[5] The struggle between Taipei and Beijing might end through peaceful unification, effectively invalidating a U.S. military response and any associated

---

[4] Sam Meredith, "Biden Says U.S. Willing to Use Force to Defend Taiwan—Prompting Backlash from China," CNBC, May 23, 2022.

[5] For examples of military analyses of the so-called Taiwan scenario, see David A. Shlapak, David T. Orletsky, Toy I. Reid, Murray Scot Tanner, and Barry Wilson, *A Question of Balance: Political Context and Military Aspects of the China-Taiwan Dispute*, RAND Corporation, MG-888-SRF, 2009; Jim Thomas, Iskander Rehman, and John Stillion, *Hard Roc 2.0: Taiwan and Deterrence Through Protraction*, Center for Strategic and Budgetary Assessments, December 2014; and Stacie Pettyjohn, Becca Wasser, and Chris Dougherty, *Dangerous Straits: Wargaming a Future Conflict over Taiwan*, Center for a New American Security, June 2022.

planning and posturing. Accordingly, the RAND Corporation's National Security Supply Chain Institute developed the *Assessing the Impacts of Interdependence* exercise to expand the conversation about China-Taiwan unification beyond the military-operational viewpoint; instead, the game grapples with the challenge from a broader, geopolitical perspective.

A New Approach to an Old Problem

Given Taiwan's crucial role in the global semiconductor supply chain, how would the peaceful annexation or outright invasion of Taiwan affect the United States, its allies and partners, and the global economy as a whole? What are Washington's options for mitigating or reversing the unfavorable effects of either peaceful or contested unification? Although these questions might provoke an uncomfortable conversation that pits the United States' economic interests against its espoused security commitments, they nevertheless deserve a sober and objective analysis, as Beijing and Washington initiate a new era of "strategic competition."[6]

Overall, the *Assessing the Impacts of Interdependence* exercise was conducted to trigger conversation and debate about the roles of the executive branch, the legislative branch, and private industry in a major supply chain disruption. Given the complexity of the crises presented and the facilitator-imposed communication challenges, we did not necessarily intend or expect to find comprehensive solutions to each problem set presented within the exercise. By fostering discussion, we aimed to inspire interaction among stakeholders and establish and improve communication about supply chain issues before a crisis occurs. By bringing together key members of each interest group, we also built the first link in a potential chain of continued interactions. As these interactions—both interpersonal and interorganizational—mature over time, they will help form the connective tissue necessary for acting quickly and decisively if a crisis ultimately does occur.

---

[6] In 2021, the Biden administration introduced the term *strategic competition* to describe the United States' complex relationship with the People's Republic of China (PRC). See Joseph R. Biden, Jr., *Interim National Security Strategic Guidance*, White House, March 2021.

Although the results of this exercise are a product of the specific individuals and groups convened to participate (which we talk more about in Chapter 3), they nevertheless highlight some important findings and some general areas in need of additional research, discussion, and gaming.

CHAPTER 2

The World Semiconductor Industry

The semiconductor industry is a very complicated market and industry sector; full consideration would go far beyond the scope of this project, and the discussion in this chapter is not intended as a comprehensive review. However, the semiconductor market offers a key lesson about the potential consequences of interdependence. The preeminence of Taiwan in the advanced semiconductors market rests squarely in the middle of a long-standing geopolitical disagreement over the future and status of Taiwan. So, although we drew heavily on existing research, we focused our analysis on how the potential for international conflict might have evolved as a result of increased economic interdependence.

Semiconductor Supply Chain Evolution

In 1958, Jack Kilby worked independently through Texas Instruments' vacation shutdown to produce the world's first integrated circuit.[1] Over the next 64 years, the semiconductor manufacturing system evolved from one man in a lab motivated by military applications (Kilby was inspired by his work

---

[1] Many scientists were working on the concept concurrently, including Robert Noyce at Fairchild and Kurt Lehovec of Sprague. See David Brock and David Laws, "The Early History of Microcircuitry: An Overview," *IEEE Annals of the History of Computing,* Vol. 34, No. 1, January–March 2012; and Jack A. Kilby, "The Integrated Circuit's Early History," *Proceedings of the IEEE,* Vol. 88, No. 1, January 2000.

on proximity fuses during World War II)[2] to a massive international industry dominated by consumer demand.[3]

Although there have been many forces driving development of the semiconductor industry, it is worth noting that, in the 1980s, there were concerns about potential Japanese dominance of the industry, which spurred a determined effort to keep some semiconductor manufacturing within the United States. These efforts resulted in the formation of a nonprofit consortium, Semiconductor Manufacturing Technology (SEMATECH), which was intended to promote semiconductor manufacturing within the United States.[4] As to whether this effort was successful, Japan did fail to become a dominant player in the semiconductor industry but, as we will discuss, so did the United States. Many factors have driven the semiconductor industry to its existing state, and it should not be assumed that government intervention will be productive.

The semiconductor supply chain has shown itself to be both very brittle and interconnected to a highly complex economic web. These connections were on clear display in March 2021 when a fire in a Renesas Electronics factory north of Tokyo effectively shut off the flow of critical automotive semiconductor chips worldwide.[5] The resulting chip shortage was blamed on the coronavirus disease 2019 (COVID-19) pandemic, but in fact, the cause of the disruption was this single event rippling through the entire global economy.

---

[2] Kilby, 2000, p. 109.

[3] Christopher Miller's *Chip Wars* was not available at the time we carried out the TTX, but it provides valuable information on the history of the semiconductor industry. See Chris Miller, *Chip War: The Fight for the World's Most Critical Technology*, Scribner, 2022.

[4] Robert D. Hof, "Lessons from Sematech," MIT Technology Review, July 25, 2011.

[5] Yang Jie, "Fire at Giant Auto-Chip Plant Fuels Supply Concerns," *Wall Street Journal*, March 23, 2021.

A Highly Connected Market

The iPad box says: "Designed by Apple in California. Assembled in China. Other items as marked thereon." Those words are not simply a label but a short summary of a highly complex supply chain spanning virtually every continent. For the semiconductor devices that form the core of today's leading-edge technology—and that are found in the cutting-edge products produced by Apple, Samsung, and others—a key supply chain node is located in Taiwan (Figure 2.1, node 4/E).

The semiconductor supply chain evolved to this state organically. Different nodes in the chain became dominant over time through locally specialized technical capabilities, education infrastructure, labor costs, strategic investment, and efficiencies of scale. These factors, along with the rise of digital design and engineering, combine with optimized logistics to make geography an almost insignificant factor.

The United States dominates multiple aspects of semiconductor production, including electronic design automation, advanced manufacturing equipment, and core intellectual property (IP) (i.e., the electronic design of the respective circuits in manufacturable format). This value is reflected in the "Designed in California" caveat on the iPad box. East Asia dominates wafer fabrication—fabricating the semiconductor chips in twelve-inch wafers, which are then tested and diced into individual chips and packaged for end-product assembly by Outsourced Semiconductor Assembly and Test (OSAT) firms (as illustrated in Figure 2.2). China is the leader in end-product assembly, packaging, and testing.[6]

---

[6] Varas et al., 2021, p. 4.

FIGURE 2.1

A Notional Supply Chain for a High-End Semiconductor Logic Chip

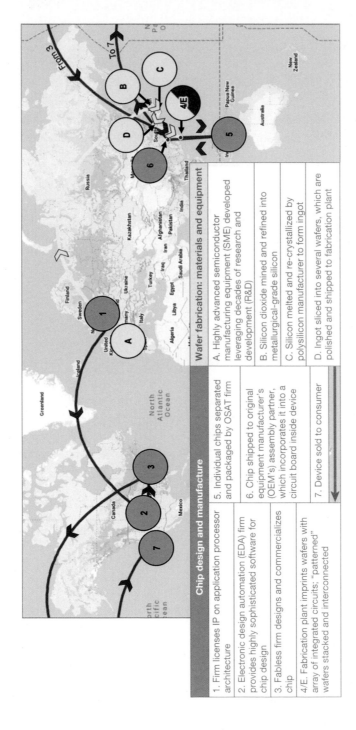

Chip design and manufacture		Wafer fabrication: materials and equipment
1. Firm licenses IP on application processor architecture	5. Individual chips separated and packaged by OSAT firm	A. Highly advanced semiconductor manufacturing equipment (SME) developed leveraging decades of research and development (R&D)
2. Electronic design automation (EDA) firm provides highly sophisticated software for chip design	6. Chip shipped to original equipment manufacturer's (OEM's) assembly partner, which incorporates it into a circuit board inside device	B. Silicon dioxide mined and refined into metallurgical-grade silicon
3. Fabless firm designs and commercializes chip	7. Device sold to consumer	C. Silicon melted and re-crystallized by polysilicon manufacturer to form ingot
4/E. Fabrication plant imprints wafers with array of integrated circuits; "patterned" wafers stacked and interconnected		D. Ingot sliced into several wafers, which are polished and shipped to fabrication plant

SOURCE: Adapted from Antonio Varas, Raj Varadarajan, Jimmy Goodrich, and Falan Yinug, *Strengthening the Global Semiconductor Supply Chain in an Uncertain Era*, Boston Consulting Group and Semiconductor Industry Association, April 2021, Exhibit 12.

FIGURE 2.2

Testing Individual Chips on a Wafer (OSAT)

SOURCE: © Can Stock Photo Inc./sspopov.

Global Foundry Manufacturing

The Taiwan Semiconductor Manufacturing Company (TSMC) provides *global foundry manufacturing*, which means they develop, build, and operate the chip fabrication facilities (commonly called *fabs*) where semiconductor chips are fabricated for other companies. For example, if a chip is designed and sold by Qualcomm, the semiconductor circuit inside the package might have been fabricated by TSMC, Samsung, or the Semiconductor Manufacturing International Corporation (SMIC). In that case, Qualcomm is a *fabless* chip manufacturer, providing key electronic components to OEMs, whose name is on the electronics that consumers and governments buy. In the Qualcomm example, most high-end Android smartphones contain Qualcomm Snapdragon chips.[7]

The five leading U.S. fabless chip manufacturers are Apple, Broadcom, Qualcomm, Nvidia, and Advanced Micro Devices (AMD). Essentially, these and other fabless chip manufacturers share the expense and risk of chip development and manufacturing with the global foundries. Fabs are extremely expensive to design, build, and equip. By assuming these burdens and leveraging economies of scale, the global foundry business model has become highly profitable and efficient. TSMC is the world's largest chip manufacturer and ranks among the most valuable companies in the world, with a market cap of over $400 billion.[8] Its dominance over cutting-edge chips comes in part from over four decades of investment.

Semiconductor device manufacturing is on track to earn $661 billion in revenue in 2022, a 13.7 percent increase over the $582 billion earned in 2021.[9] As shown in Table 2.1, the top five U.S.–based fabless chip firms account for $345 billion in sales (2021) and nearly $3 trillion in market capitalization.

With 92 percent of market share, TSMC dominates the global foundry production of cutting-edge semiconductors (i.e., those with circuit device resolution less than ten nanometers [nm]).[10] Lower resolution in wafer fabrication translates to faster speeds and improved energy consumption. These cutting-edge chips drive most new mobile phones, smart watches, and cutting-edge commercial, industrial, and military electronics. TSMC is scheduled to start delivering 3 nm resolution wafers this year (2022)—a capability that places TSMC above its nearest competitor, Samsung.[11] TSMC accounts for 54 percent of the global foundry market share, and Taiwan

---

[7] Teejay Boris, "Qualcomm Snapdragon Chips Mostly Power $300+ Smartphones | MediaTek Tops Cheaper Devices," *Tech Times*, March 21, 2022.

[8] Companies Market Cap, "Largest Companies by Market Cap," webpage, undated.

[9] International Data Corporation, "Worldwide Semiconductor Revenue to Grow 13.7%, but Supply Chain Remains Selectively Challenging Amidst Global Economic Volatility, According to IDC," webpage, June 8, 2022.

[10] Varas et al., 2021, Exhibit 17.

[11] Govind Bhutada, "The Top 10 Semiconductor Companies by Market Share," *Visual Capitalist*, December 14, 2021.

TABLE 2.1
Top Seven Fabless Manufacturers

Fabless OEM Firms	Annual Sales ($ billion)			Market Cap July 2022 ($ billion)	Applications (% of 2018 Sales)	Main Products (% of Market Share)	Key End-Market Segments	Key Foundry Providers
	2010	2018	2021					
Broadcom (U.S.)	6.7	17.5	27.5	198.5	Logic (79)	Wireless LAN chips (50 to 60)	Mobile phones, PCs	TSMC (main), GlobalFoundries, UMC
Qualcomm (U.S.)	7.2	16.6	33.6	158.8	Logic (71)	Mobile processors (40), baseband chips (60)	Mobile phones, PCs, consumer electronics	TSMC, Samsung, SMIC
Nvidia (U.S.)	3.1	10.4	26	65	Logic (100)	Graphics chips (80)	PCs, automotive	TSMC (main), Samsung
MediaTek (Taiwan)	3.5	7.9	—	—	Logic (90)	Mobile processors, wireless LAN chips	Mobile phones, consumer electronics	TSMC (main), UMC, GlobalFoundries
Apple (U.S.)	65	166	242	2,390	Logic (100)	Mobile processors (Apple)	Mobile phones (Apple)	TSMC (main > 2016) Samsung (main < 2016)

Table 2.1—Continued

Fabless OEM Firms	Annual Sales ($ billion)			Market Cap July 2022 ($ billion)	Applications (% of 2018 Sales)	Main Products (% of Market Share)	Key End-Market Segments	Key Foundry Providers
	2010	2018	2021					
AMD (U.S.)	6.4	6	16.4	132.5	Microprocessors (63), logic (37)	Graphics chips (10), microprocessors (5)	PCs, consumer electronics	TSMC (main > 2020) GlobalFoundries (main < 2020)
HiSilicon (China)	0.3	5.5	—	—	Logic (100)	Mobile processors (Huawei)	Mobile phones (Huawei)	TSMC (main), SMIC
U.S. totals	88.4	216.5	345.5	2,944.8				

SOURCES: Henry Wai-Chung Yeung, "Explaining Geographic Shifts of Chip Making Toward East Asia and Market Dynamics in Semiconcuctor Global Production Networks," *Economic Geography*, Vol. 98, No. 3, 2022, p. 292; and Google Finance, homepage, undated.

NOTE: PC = personal computer; UMC = United Microelectronics Corporation.

accounts for a total of 63 percent (Table 2.2) and 41 percent of all wafer fabrication.

Another key factor that has led to TSMC's industry dominance is customer intimacy. On the one hand, fabless firms need to establish a market and OEM demand for their chips by working closely through the design process with the OEMs. On the other hand, fabless firm chip orders can be properly fulfilled only if the firms have strong foundry support and guaranteed capacity allocation. In turn, trusted foundry providers can reciprocate this customer intimacy through investment in new cutting-edge fabs, capacity, process innovation, and customization.[12]

TABLE 2.2

Global Foundry Companies by Market Share and Country

Company	Market Share (%)	Country
TSMC	54	Taiwan
Samsung	17	South Korea
UMC	7	Taiwan
GlobalFoundries	7	United States
SMIC	5	China
HH Grace	1	China
PSMC	1	Taiwan
VIS	1	Taiwan
DB HiTek	1	China
Tower Semiconductor	1	Israel
Other firms	5	N/A

SOURCE: Bhutada, 2021.

NOTE: N/A = not applicable; PSMC = Powerchip Semiconductor Manufacturing Corporation; VIS = Vanguard International Semiconductor Corporation.

---

[12] Yeung, 2022, pp. 291–293.

A recent example of this intimacy between fabless manufacturers, global foundries, and OEMs is Qualcomm's new 4 nm smartwatch chip, designed specifically for the next generation Google Wear operating system (Figure 2.3). This new product reflects intimate cooperation among Qualcomm, the fabless firm whose name will go on the outside of the chip; TSMC, the global foundry providing new chip fabrication methods and technology; Google, who designed the operating system; and several OEMs who will package and market the commercial end product to the public.

All these entities worked together on the project from its conception. Qualcomm will leverage the 4 nm chip manufacturing process and fab infrastructure developed by TSMC, an improvement over the previous 12 nm chip process, to provide a watch logic chip for Google Wear operating system specifications that operates twice as fast and provides 50 percent more battery life—specifications demanded by the commercial market-

FIGURE 2.3
Smartwatch Logic Processor

SOURCE: Qualcomm promotional image.

place. The OEMs (Google does not manufacture watches) will be Chinese companies, such as Oppo and Mobvoi.[13]

Product development at this level leads to highly interdependent nodes in the supply chain and highly specialized products. Leading-edge consumer products require an integrated development process that produces logic chips with very specific applications. These products do not interchange and have no shelf value other than for the specific end-product for which they were designed; a Qualcomm chip for a Google Wear device does not interchange with an Apple chip for an Apple Watch or vice versa.

Where certain types of chips, such as memory chips, might be interchangeable as commodities between products, they tend to be ancillary and do not form the core of the product design that gives the product most of its value. As a practical example, one can plug a SanDisk or a Vansuny flash drive memory card into a laptop or camera interchangeably, but the logic and the engineering inside the laptop, camera, or other device is what gives that memory utility and provides the value to the user.

The supply chain for such leading-edge products develops over the course of the engineering lifecycle. As in the Google Wear example, Qualcomm, TSMC, Google, and the OEMs have built a relationship through the development process that ultimately results in a commercial product supply chain to optimize revenue. A wafer fab for a specific leading-edge product cannot readily be changed to make a different product. To remove or change a supply chain node likely requires significant investment and time.

This supply chain relies heavily on refined customization at each node, developed in concert to optimize for specific consumer products. Within the cutting-edge wafer fab, the key manufacturing elements have been designed, refined, and programmed for specific products. Behind the TSMC fab design is a supply chain of equipment manufacturers who work in concert with TSMC to advance research and push the state of the art. Thus, the complexity and specifics of the supply chain make it brittle.

The reasons that TSMC has attained a dominant position in the semiconductor market are not nefarious but are largely driven by the market itself. TSMC is technically proficient, operates in a capital-intensive portion

---

[13] Julian Chokkattu, "Qualcomm's New Smartwatch Chips Promise Big Battery Life Gains," *Wired*, July 19, 2022.

of the supply chain, receives dedicated support from the Taiwanese government, and pursued a business model in which it operates as a foundry with multiple customers (rather than the vertically integrated model of Intel). Changing this model to expand suppliers of high-end semiconductor chips will not come about organically—it will take time and resources and will require government intervention.

CHAPTER 3

The Game and the Scenarios

We developed several objectives for the *Assessing the Impacts of Interdependence* exercise to approach the issue of Taiwanese sovereignty from a largely untrodden perspective. First, we used two scenarios in the TTX to investigate the economic and diplomatic hurdles associated with the loss of Taiwanese autonomy given Taipei's preeminence in the global production of semiconductors. While a peaceful unification scenario would award the PRC with Taiwan's manufacturing capacity, a contested unification scenario would deprive the world market of all Taiwanese semiconductor output. In either case, the economic and security dynamics of the international system would experience a jarring reshuffle, which largely has been overlooked by U.S. policymakers to date.

Second, the TTX was designed to stimulate conversation between the executive and legislative branches regarding ownership of supply chain problems. Washington must determine how it would respond—through executive action, new legislation, or both—to resolve a supply chain crisis that traverses all domains of national power. Finally, the TTX was intended to spur dialogue between government and private industry by allowing each to articulate its interests and observe where these interests converge and diverge. Profit-seeking firms might advocate courses of action that differ or even conflict with those promoted by government officials concerned about U.S. security credibility. By stressing public-private relations to such an extreme, the exercise was intended to provide a forum where both sides could discuss response options before either hypothetical scenario becomes an irreversible reality.

Game Methodology and Structure

In alignment with its core objectives, the RAND exercise relied on the input of professionals from the U.S. executive branch of government, the legislative branch, and private industry to evaluate each scenario. Collectively, these three sub-teams formed a Blue cell tasked with responding to challenges presented by RAND facilitators, the White cell.

The primary purpose of the exercise was to show Blue players the challenges of the environment, not necessarily to match wits with a reacting adversary. Consequently, the exercise did not use a Red cell player acting as the PRC. Instead, the White cell depicted Chinese actions as injectors into the overall gameplay. The RAND team used a similar approach for capturing the behavior of the Yellow cell, representing Taiwan, and the Green cell, representing the rest of the world. In future games, separately played Red, Yellow, and Green cells might be desirable, but for this game, whose primary focus was defining the problem, White cell inputs were sufficient.

After reviewing the key characteristics of each scenario, the White cell physically separated the executive, legislative, and industry sub-teams to emulate siloed communications. Once the sub-teams were formed, the White cell facilitators presented each group with a decision tree visualization tailored to the group's particular capabilities and organizational interests. This diagram also outlined the scripted actions leading from the start of the exercise to certain decision points in the game.

At each decision point, the Blue sub-teams were presented with options for addressing the unfolding crisis. These options were merely suggestions for responding to the gameplay; they were intended to stimulate discussion and did not serve as a restrictive menu from which players had to select an action. Once each Blue sub-team agreed on its course of action, the entire Blue cell reconvened. A representative from each sub-team then summarized the group's findings in a briefing to the full TTX enterprise. Once the three sub-teams shared their decisions, the White cell gathered separately for adjudication to determine whether the Blue cell's actions ultimately succeeded or failed in realizing their intended effects.

Common Conditions Between Scenarios

Both of the scenarios—peaceful unification and contested unification—begin from a common starting point.[1] According to 2021 data, Taiwanese fabs manufacture 41 percent of the world's high-end logic chips.[2] These tiny devices serve as the brains of modern electronics because of their abilities to perform high-speed processing and computation. They are critical components of modern technologies, ranging from cell phones to artificial intelligence systems. Combined, the United States and Europe produce 44 percent of the world's high-end logic chips, while the PRC manufactures just 2 percent. In terms of the IP required to design and build high-end semiconductors, the United States and Europe account for 94 percent compared with the PRC's 4 percent (Figure 3.1). Taiwan clearly dominates the physical fabrication of semiconductor chips (as the previous chapter described), but the United States and Europe overwhelming provide the unique knowledge necessary to produce and advance semiconductor technology.

In Figure 3.1, the geographic share in each step of the semiconductor supply chain under the status quo is portrayed for chips of resolution less than 22 nm. From left to right, the United States—and to a lesser extent, Europe—dominate the IP and design of these high-end chips. Designing and manufacturing the cutting-edge equipment required to produce high-end chips is dominated by the United States, Japan, and Europe, with key equipment supplied from the Netherlands. The provision of raw materials is not dominated by a single supplier, but more than half come from Asian countries, including Chinese-owned mining and refining entities. Taiwan and the United States dominate wafer fabrication for devices smaller than

---

[1] The scenarios chosen for this TTX are not the only sources of potential disruption to semiconductor supply. Other disruptions worthy of study include a targeted cyberattack by the PRC on the semiconductor industry; a physical or cyberattack by the PRC on the undersea cables to Taiwan (effectively an information quarantine or blockade); and the ability of remote actors, such as the companies that make the manufacturing equipment, to render the semiconductor fab equipment inoperable. The dependence of the semiconductor industry on electronic communications for design, manufacturing, and supply chain management renders it vulnerable to a number of cyberattack media.

[2] Varas et al., 2021, Exhibit 17. For the purposes of this exercise, a *high-end* logic semiconductor refers to those that are smaller than 22 nanometers in diameter.

FIGURE 3.1

The High-End Logic Semiconductor Supply Chain Under Status Quo Conditions

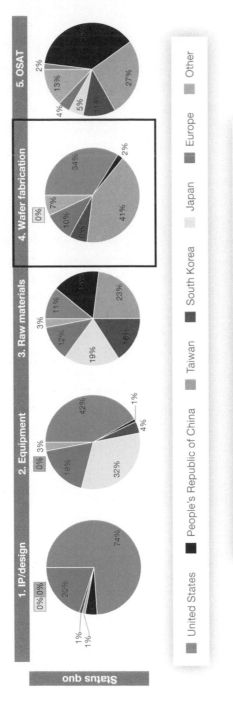

Key points:

1. United States and Europe own 94 percent of all high-end logic chip IP.
2. Taiwan's fabrication plants produce 41 percent of the world's high-end logic chips.

SOURCE: Adapted from Varas et al., 2021.

NOTE: Europe in this figure primarily consists of Germany, the Netherlands, and the United Kingdom.

22 nm, but for devices smaller than 10 nm, Taiwan has over 90 percent of world capacity.[3] Firms in China, Taiwan, South Korea, and Malaysia dominate OSAT, as described in Chapter 2.

In 2025, an ascendent PRC demands that Taiwan cease autonomous governance and accept political and economic unification with the mainland. Beijing subsequently deploys military and paramilitary forces to interdict and inspect all air and sea cargo entering and exiting the island through what the Chinese Communist Party (CCP) calls a "quarantine" of the PRC's most unruly province.[4] Although the inspection regime does not significantly hinder the flow of goods to or from the island, Beijing does explicitly prohibit Taiwan from exporting high-end semiconductors to the United States. Senior leaders within the Central Military Commission push for this policy because they fear that Washington will use the chips to develop next-generation weaponry that could threaten Beijing's geopolitical designs for Southeast Asia.

Beijing's unification stipulations leave the Taiwanese population deeply divided and conflicted. On one hand, pragmatists argue that Taiwan must adapt to the changing geopolitical environment and accept a favorable reconciliation with the PRC while such an opportunity still exists. On the other hand, nationalists espouse the principle of Taiwanese independence, especially because younger generations have grown more socially and culturally distant from their mainland cousins. It is within this context that Taipei must decide whether it will continue the path of peaceful unification, or reverse course and contest subordination to the mainland.

Scenario 1: Peaceful Unification

Fearful that the PRC might tighten the noose of the quarantine, the Taiwanese people debate their options throughout 2025. After a year of national deliberation, Taipei's leaders agree to peacefully unify with the mainland in

---

[3] Varas et al., 2021, Exhibit 17.

[4] This scenario is based on the one outlined in Bradley Martin, Kristen Gunness, Paul DeLuca, and Melissa Shostak, *Implications of a Coercive Quarantine of Taiwan by the People's Republic of China*, RAND Corporation, RR-A1279-1, 2022.

a clear sacrifice of economic and political freedom for long-term security. Once both sides sign the unification agreement, People's Liberation Army (PLA) forces occupy the island, and Beijing sends CCP officials to inspect the facilities of TSMC. Without interrupting production, the PRC steadily assumes control of the fabs, as well as the highly skilled labor force that operates them. With this bloodless acquisition of Taiwan's fabs, the PRC now produces 43 percent of the world's high-end logic semiconductors, including over 90 percent of the most advanced chips.[5] Beijing subsequently dominates two critical nodes in the semiconductor supply chain—silicon wafer fabrication and OSAT (Figure 3.2)—allowing the PRC to manipulate supply and set market prices.

Domestically, the CCP soon imposes the draconian social and economic policies of the mainland on the Taiwanese people. Discontented with this new authoritarian rule, large numbers of highly skilled, technically trained Taiwanese professionals flee their homeland. Relationships between U.S. and Taiwanese companies also begin to sour, and international IP arrangements between Taiwan and the West evaporate. These developments lead economic analysts to question the PRC's ability to maintain pre-unification levels of Taiwanese semiconductor output. Although China has gained physical control of Taiwan's foundries, it might be losing access to critical inputs—IP and skilled labor—at higher sections of the supply chain.

Gameplay for Scenario 1

The Blue sub-teams were divided to reflect the perspectives of the executive and legislative branches of the federal government and the views of U.S.-based industry. In the peaceful unification scenario, RAND facilitators prompted each of these groups for a course of action at a single decision point: the departure of Taiwanese skilled workers from their homeland. Specifically, RAND facilitators asked the executive branch team to consider such actions as restricting the flow of IP to Taiwanese fabs, encouraging

---

[5] Semiconductors smaller than ten nanometers in diameter are considered the most advanced in the world. As of 2022, the smallest and most powerful chips have three-nanometer diameters. See TSMC, "Logic Technology," webpage, undated.

FIGURE 3.2

The High-End Logic Semiconductor Supply Chain: Status Quo Versus Peaceful Unification Conditions

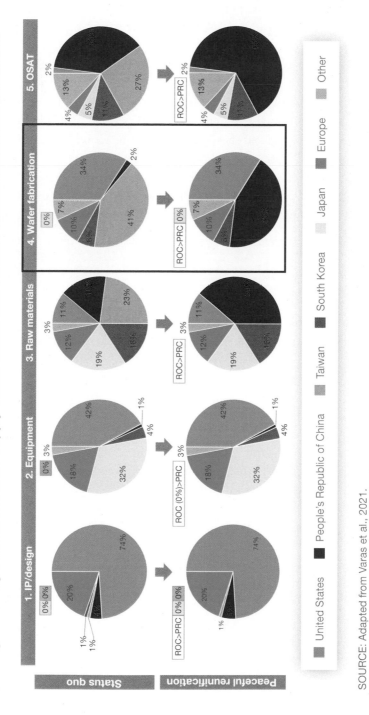

SOURCE: Adapted from Varas et al., 2021.
NOTE: The red box around the column indicates a notably decisive impact. ROC = Republic of China (Taiwan).

allies and partners to cease collaboration with the PRC, and developing new immigration policies to attract skilled Taiwanese emigres.

Each option would require input from key executive agencies, such as the Department of Homeland Security and Department of Commerce. For the legislative branch, response options might include drafting new immigration reform legislation or extending existing sanctions against the PRC to occupied Taiwan. The wide-ranging effects of peaceful unification—from economic to diplomatic and social—would require action by multiple committees within both houses of Congress. Industry's response options would vary significantly from the executive and legislative branch actors; it may continue business as usual with the PRC, despite protest from the federal government, or it may comply with Washington's requests, at the expense of profit.

Scenario 2: Contested Unification

Despite the looming threat of a more restrictive quarantine—and possibly a full-scale Chinese invasion—Taiwan boldly rejects unification with the PRC roughly one year after receiving Beijing's ultimatum. Taipei's defiance, however, does not necessarily signal an intent to resolve the crisis militarily. Still hopeful that diplomacy can prevail, Taiwanese leaders declare that the island will not attempt to break the Chinese quarantine with force. They command their military forces to stand down and ask the United States to refrain from intervening in any way. To signal to Beijing its desire to resolve the crisis peacefully, Taipei also expels all foreign troops from Taiwanese territory.[6]

Economically, the Taiwanese government adopts a *middle-ground* approach with respect to the exportation of high-end logic semiconductors. Because China has demanded that Taiwan cease selling its most advanced chips to the United States, Taipei believes that it can appease Beijing and minimize damage to its relations with Washington by declaring a halt to *all*

---

[6] Benjamin Angel Chang covers a similar contested scenario in *Artificial Intelligence and the US-China Balance of Power*, dissertation, Massachusetts Institute of Technology, June 2021.

high-end logic chip exports (Figure 3.3). The entire world market has now been deprived of access to high-end Taiwanese semiconductors, including those smaller than ten nanometers in diameter—the cutting edge of semiconductor technology.

Outraged by Taipei's decision, Beijing tightens the quarantine and reduces Taiwan's fuel imports to a subsistence trickle. Basic services on the island remain uninterrupted, but large-scale factory production abruptly halts, and aggregate semiconductor output—to include both low-end and high-end chips—falls to zero. Denied access to all forms of Taiwanese semiconductor technology, the global market contracts and major economies begin to collapse. Although the PRC could easily reverse this disruption by restoring Taiwanese fuel imports, Taipei threatens to destroy its fabs and permanently disrupt the global supply chain should the PLA attempt a full-scale invasion. The loss of Taiwanese production capacity would generate widespread economic calamity that would endure until alternative manufacturing capabilities could be identified, resourced, and constructed. Accordingly, a rapid resolution to this crisis remains in the best interest of the United States, the PRC, and the broader world economy.

Gameplay for Scenario 2

A contested unification scenario presented the legislative, executive, and industry teams with two sequential decision points. First, the executive branch, legislative branch, and industry teams must determine how they would respond once Taiwan hedges against the PRC and the United States by halting all global high-end semiconductor exports. The executive and legislative branch teams might, as one example, assume the role of mediator and adjudicate how the existing supply of high-end chips should be allocated across the U.S. economy. Or, these teams might implement a policy designed to block Chinese acquisition of raw materials, thereby inhibiting the PRC's ability to produce domestic alternatives to Taiwanese fabs. Industry's options are more limited in this first decision point, given Taiwan's lack of warning in halting all high-end chip exports.

Each team encountered a second decision point in this scenario once the PRC restricted Taiwanese fuel imports, effectively shuttering the island's production of both low-end and high-end semiconductors. In this case, the

FIGURE 3.3

The High-End Logic Semiconductor Supply Chain: Status Quo Versus Contested Unification Conditions

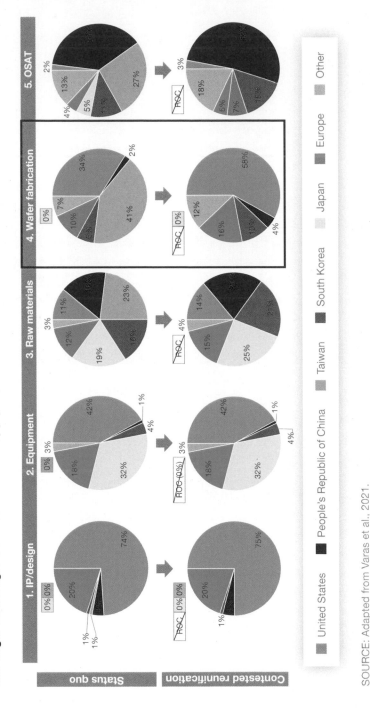

SOURCE: Adapted from Varas et al., 2021.
NOTE: The red box around the column indicates a notably decisive impact. The red slashed line indicates that Taiwan is no longer available to provide exports.

executive and legislative teams could resolve to ignore Taiwan's diplomatic pleas and break the Chinese quarantine with force. If these teams did not find the latter course of action acceptable, they might simply accept the reality of a permanently disrupted supply chain and encourage the country to modify its semiconductor consumption habits. Private industry might opt to invest in alternative production capabilities in such locations as South Korea, Africa, or South America at this decision point; alternatively, it might covertly bypass political and legal restrictions for the sake of profit. Regardless of the actions selected, each team had to contend with the loss of 41 percent of the world's high-end logic chip and 35 percent of its low-end logic chip manufacturing capacity.[7]

---

[7] Taiwan's share of low-end chip production capacity derived from data in Varas et al., 2021.

CHAPTER 4

A Vulnerable Supply Chain

At the end of both scenarios in the TTX, the U.S. government had only bad choices from which to choose, all involving either some level of accommodation to the PRC or the acceptance of economic disruption to a greater or lesser degree. Choices became stark: (1) accept the PRC's demands and, in doing so, decline to support Taiwan's claims of autonomy; (2) take actions to offset the loss of semiconductor access, which would likely lead to long-term decline in economic output; or (3) resort to an armed response to protect Taiwan and coerce the PRC to stop its quest for unification, which also would be expected to lead to global economic disruption. In the TTX's design, we excluded an armed response. Better response choices would have required action taken long before the crisis to establish some level of production capability on U.S. or allied territory.

As game players cast about for feasible responses, it became starkly clear that the vulnerabilities generated by Taiwan's dominance of the semiconductor supply chain are not well understood. Indeed, Taiwan itself might fail to understand how its development of market dominance in semiconductors has created greater vulnerability, not greater security. The PRC retains considerable leverage over Taiwan regardless of Taiwan's position in the semiconductor market, and the PRC both gains leverage over the rest of the world and secures its own security by absorbing Taiwan and its industry. Such action also puts the PRC's supply of semiconductors at risk, but the question then becomes one of relative damage.

Government Actors Generally Lacked Information on the National Implications of Semiconductor Chip Supply

Although experts on semiconductor markets and technology are no doubt present within the U.S. executive branch and on congressional committee staff, both executive and legislative branch players admitted that they did not possess sufficient information to confidently decide on a course of action in the crisis. The degree of dependency on Taiwan was surprising to many, and the first impulse of players on both sides was to ask for more information. Players called for analysis of

- supply chains
- risks to national security, critical infrastructure, consumer products, and more from the loss of fabrication capability
- options to mitigate risks of loss of access (stockpiles, recycling, and building new fabs).

In some cases, the immediate reaction of the actor was to search for military responses, even in the first scenario, in which Taiwan yielded to PRC demands without contest.

Executive and legislative actors suggested stockpiling or recycling older chips as possible solutions. Stockpiling is not realistic for high-end semiconductor chips because firms build and design these chips for very specific purposes, as explained in Chapter 2. Thus, there is no *general* type of semiconductor that can be collected en masse, which makes it impractical to stockpile semiconductors in anticipation of a global interruption. Similarly, cleaning or recycling chips are also impractical solutions largely because of the specific use for which chips are designed or the need for fabs to perform actual production. These complications were neither known nor considered by the players prior to being proposed and assessed in the TTX.

Equally impractical are solutions that require implementation timeframes that extend far beyond the expected duration of the crisis. A semiconductor fab is a complicated factory with a large concentration of equipment and a complement of highly skilled personnel. Building a fab is a years-long process; creating a workforce to man a fab might take a generation. Attempting to respond to an immediate crisis with a process that will take years to implement is not practical. The likely impact would be market disruption serious enough to create a depression.

Having taken military action off the table for the TTX, the United States and its allies are left with the choice of either capitulating to PRC demands and possibly abandoning Taiwanese claims of autonomy, or accepting years-long dislocation and diminished economic growth. The starkness of this choice was not generally understood.

Government and Industry Interests Are Likely to Diverge

U.S. industry has devolved many of its functions overseas for solid business reasons: A skilled labor force is available, as are well-established companies with a track record of timely delivery. Absent disruption to the supply chain, companies have every reason to maintain existing relationships. Their incentives are different from government actors and motivated by market and shareholder value.

In the peaceful unification scenario, industry players sought to continue business as usual. They did not perceive major risks from a takeover of Taiwanese industry by the CCP, at least in the short term. Industry players indicated that they viewed measures intended to exert leverage over the PRC—such as restricting IP and encouraging skilled labor to emigrate from Taiwan—as counterproductive. For example, industry had no interest in changing immigration policy to allow easier immigration from Taiwan, believing it better for industries to keep a stable employment base in Taiwan.

Overall, in the peaceful unification scenarios, industry viewed U.S. government action as the major risk to its business—specifically, that action to restrict access to semiconductor manufacturing in a unified Taiwan or to encourage immigration would promote market and labor instability. Industry representatives did believe that over the longer term they would need to seek other supply sources, but they viewed most immediate actions by the government as precipitous and counterproductive.

Industry did see a role for government in the contested unification scenario, in which major economic disruption and shortages of key material supplies would be likely. But the solutions—effectively, the government taking over semiconductor distribution in the United States until alternatives could be found—were at a level of interference that have not been seen in the United States since World War II. Extensive government intervention implies that the supply of key commodities would have become so diminished that government rationing was the only practical alternative for ensuring supply to the nation's highest priorities.

The Degree of Allied and Partner Interest in Semiconductor Supply Was Not Fully Explored

Semiconductors cross borders multiple times during fabrication and installation into end items, to the point that multiple actors in multiple countries have a stake in the overall process. Action by any one country or actor could potentially disrupt the market; actions by multiple actors would be needed to secure the market. The TTX structure allowed actors to consider allied and partner interests but did not specifically include allied and partner players. This limitation probably constrained realistic dialogue but might not have materially changed the available responses to Chinese actions. The needs of the European Union to gain access to semiconductors likely make the choice between capitulation or economic disruption even more stark. It is possible that U.S. allies offer some alternative sources of chip manufacturing, but they might be influenced in different ways by U.S. choices. These relationships are important because of the global nature of the supply chain and deserve a more detailed examination; however, they do little to change near-term options that are uniformly undesirable.

Increasing the Level of Understanding and Prior Planning Is an Important First Step

The scenarios offered in the exercise were neither unrealistic nor projected to occur in the far future. But, the implications of the scenarios were only beginning to be understood by the players, and the available solutions were unlikely to be effective. The need to completely analyze national security and broader economic implications and provide broader awareness of the challenges and timelines for action was a major outcome of the TTX. In short, it was evident that if this crisis occurred tomorrow, the United States would not be prepared. Having better options requires prior planning and awareness that some potential courses of action have long timelines.

The Biden administration has taken the initiative to improve U.S. competitiveness in chips manufacturing, specifically the Creating Helpful Incentives to Produce Semiconductors for America (CHIPS) Act.[1] This legislation was being considered by Congress during the TTX and was viewed by participants as at least helpful in advancing the goal of reduced dependency on semiconductor production in East Asia.

However, the exact impact—if there will be any impact—on the choices available in a crisis was not known, and the $39 billion of incentives contained in the CHIPS Act does not put a significant dent in the $661 billion semiconductor market. Although the initiative provides incentives for companies to invest in U.S. infrastructure, it does not come close to reducing vulnerability in a way that would afford more or better choices in the short term.

---

[1] National Institute for Standards and Technology, "CHIPS Act," webpage, updated April 5, 2022.

CHAPTER 5

Economic Consequences

The TTX and the analysis to support it did not include a detailed econometric analysis of sector-by-sector impact. However, we can say with some confidence that *without preparation, a major shift in semiconductor supply would have an immediate and significant impact on the U.S. economy overall; the outcomes of the second scenario would be particularly dire.* Taking national security into account more broadly than military posture and readiness, lack of access to semiconductor chips would have an impact on prosperity and well-being and could even have an impact on overall health and welfare. These impacts would be significant even in the absence of outright military action.

Semiconductors Touch Almost Every Sector and Large Numbers of Products

The systems that provide us with cutting-edge consumer electronics, artificial intelligence, and gas pumps with video screens include a common hardware element: semiconductors. Defining a system as "a set of elements so interconnected as to aid in driving toward a common goal,"[1] the system that produces semiconductors contains scores of elements—from raw materials and supply chains to microscopic device fabrication and product assembly—all driving toward the common goals of profit-making and maintaining market share. Figure 5.1 depicts, at a very general level, the places where TSMC, the world's dominant chip maker, based in Taiwan, affects the world economy.

---

[1] John E. Gibson, William T. Scherer, William F. Gibson, and Michael C. Smith, *How to Do Systems Analysis: Primer and Casebook*, Wiley, August 1, 2016, p. 3.

FIGURE 5.1
Market Impacts of Semiconductors

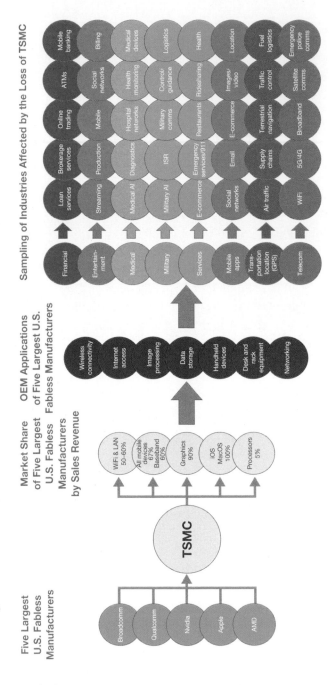

SOURCES: Yeung, 2022, p. 292; Google Finance, homepage, undated; Brian Dean, "iPhone Users and Sales Stats for 2022," webpage, Backlinko, updated May 28, 2021.

NOTE: AI = artificial intelligence; ATM = automated teller machine; comms = communications; GPS = Global Positioning System; ISR = intelligence, surveillance, and reconnaissance; LAN = local area network.

This economic ecosystem relies on technology pushing the leading edge of semiconductors, while the previous generations of memory and ancillary chips became commodities (see Chapter 2). Chips went from being expensive, bespoke devices that only the military could afford to being so common that many of us carry devices driven by cutting-edge chips in our pockets and on our wrists on a daily basis. The backbone networks and cloud services that invisibly provide those devices with their capabilities also depend on cutting-edge chips, as do AI, satellite technologies, and certain advanced military capabilities. As semiconductors have grown more powerful, efficient, and omnipresent, it might be hard to find a facet of the U.S. economy that is not affected by the system that produces them.

Economic Impacts Could Range from Concerning to Devastating

Interconnected economies create vulnerability if such factors as geopolitics result in a disruption to the supply chain. If semiconductors are denied to tech industries, these industries can no longer count on continued technological improvement as a means to maintain growth and market share. If semiconductors are unavailable to other industries, production suffers and shortages develop, as seen in the commodity chips shortage experienced in the auto industry in 2021.[2]

This disruption would affect both the PRC and Western economies. Even if the PRC managed to completely secure the supply of chips, economic disruption in the rest of the world would lower demand for Chinese goods, meaning that fewer consumers globally would have the means to consume goods produced by PRC companies. Every major economy would suffer, leading to the following relevant questions:

1. Who would suffer more and most immediately?
2. Who would best be able to adjust and overcome the disruption?

---

[2] Michael Wayland, "Chip Shortage Expected to Cost Auto Industry $210 Billion in Revenue in 2021," CNBC, September 23, 2021.

This adjustment would involve more than relocating production; it would involve changing, in basic ways, how the supply chains and markets would function.

Such a change would require time, capital, an available workforce, and possibly improvements in technology. We know that it would take two to five years for the United States and its allies to build and outfit sufficient fabrication capacity to offset the loss of Taiwan's production. And this timeline is based on optimistic assumptions about tooling, permitting, and the labor market. In contrast, China—an autocratic society better able to harness the whole of government and the economy to pursue objectives—could probably build infrastructure faster, but we have not done the analysis to assess how quickly the PRC could replicate Western tooling or develop the required labor market. The issue might be less about how quickly the PRC could generate new capacity than how long it could stand the overall decline in global economic activity.

Industry Perspectives Focused on Maintaining Relationships Where Possible

Industry players in the TTX were clear that many of them routinely do business with organizations in the PRC, and a change in Taiwan's orientation might not immediately affect their business operations. Although there is no doubt that the PRC would gain economic, political, and diplomatic leverage and that this might affect business practices over time, in the short term, it would be in everyone's interest to continue existing relationships with respect to producing and consuming semiconductor chips.

However, an interruption in the supply of chips—whether because of a contested scenario or as an aftermath of the PRC achieving market dominance—would have an immediate and possibly severe economic impact. If only the highest-end semiconductors are affected, the impact would be a bifurcated market, in which the PRC would have all the production capability but the United States and its allies likely would retain design capabilities. As a consequence of such an arrangement, new and advanced semiconductors would be available to no one, with some obvious implications for productivity and a level of shared pain across world economies.

In the more severe case, if Taiwan's complete semiconductor manufacturing capacity is lost, the economic pain would be much greater, particularly in the West. Supplies of semiconductors would quickly become unavailable, resulting in extreme competition for the remaining supply—literally a commercial life-or-death situation for companies whose products rely on semiconductors.

Prices for remaining semiconductors would increase. The supply of end products that rely on semiconductors would contract in the short run. Prices for products from companies that are able to obtain semiconductors would rise significantly, both from increased production costs and from excess demand. The macroeconomic impacts would include

- developed economies experiencing inflation as many prices rise
- large increases in unemployment because many businesses that require semiconductors for their products are not able to obtain them
- a possible years-long economic depression.

In the short run, the standard of living in developed economies would be significantly degraded. Over the long run, new capacity to produce semiconductors would eventually come online, but that would be a multiyear process costing billions, as indicated earlier. Furthermore, at a time of widespread economic downturn, it would be even more challenging to bring new fabrication capacity online since it is a high-capital endeavor. Government investment becomes much harder as debt rises.

It could be an extended period before the global economy returns to its previous level of health and innovation. Thus, the loss of semiconductor fabrication capacity or loss of confidence in capacity in Taiwan would likely result in global depression for an extended period.

Realigning Industry Incentives Would Be Very Challenging

Private companies in the United States are neither expected nor structured to consider collective interest as they make business decisions. Their actions do not reflect a lack of patriotism or public spirit. Corporate leaders are responsible for the interests of their companies and often are not even in a position to understand or act on broader public goals. The existing supply chain evolved to promote efficiency, and it accordingly has put a focus on locating its components where they are most readily produced.

For the most part, the interconnections work well, with steady improvement in technology and effective delivery to consumers. The existing situation reflects years of private sector decisionmaking focusing on market forces and shareholder value, aided by manufacturing, management, and logistics approaches that have continually refined and optimized the system. Lean manufacturing, six-sigma processes, just-in-time supply chains, statistical analysis in operations, and a litany of managerial processes continuously add to ever-improving efficiency and lack of redundancy in semiconductor manufacturing.

That collective and individual interests diverge is nothing new. But the potential international environment we have described as an outcome of Taiwan unification is unprecedented and likely would require changes to how semiconductor companies manage their supply chains. On the one hand, centralized planning—the government attempting to direct distribution by administrative fiat—does not have a good history, and none of the players in the TTX viewed this as a good solution. On the other hand, there simply is no reason to believe that the market will, on its own, reduce the dependency on Taiwan for semiconductor production, a dependency that leads to geopolitical vulnerability for the United States and its allies.

As of this writing in 2022, there have been efforts to encourage U.S. companies to acquire the capability and capacity to produce high-end computer chips, most notably in the form of the CHIPS Act. These efforts were not specifically considered in the TTX, and the players had no specific recommendations other than to study the subject more thoroughly. Yet, two high-profile efforts to *reshore* production—to bring manufacturing capability back to the United States—have so far been unsuccessful. The effort to open a Foxconn manufacturing facility in Wisconsin has not yet resulted in a working factory; notwithstanding the plan for $4 billion in state incentives,[3] the facility now sits largely abandoned.

---

[3] David Shepardson and Karen Pierog, "Foxconn Mostly Abandons $10 Billion Wisconsin Project Touted by Trump," Reuters, April 21, 2021.

In another case, TSMC is encountering problems with the workforce available in Arizona. Despite federal and state subsidies and accommodation in the CHIPS Act, TSMC is now importing high-school graduate labor from Taiwan to staff the effort.[4] Moreover, there is more to the semiconductor process than fabrication. If semiconductor fabrication is to be the subject of policy and national investment, then our capabilities in other aspects of the system, such as packaging technologies in support of semiconductor manufacturing, will also be critical.

A major limitation on attempts at reshoring is lack of available labor in the United States. Although the United States has a robust collection of very capable designers and scientists, it does not have a ready source of skilled technical labor that would be needed to man the large numbers of positions that would be available in fabrication facilities. In the immediate term, there is no way to quickly create thousands of trained technical workers from the domestic labor market. If new labor is required, the most likely source will be immigration.

However, resolving labor shortages must keep pace with the other elements required to create the facilities. Constructing new fabs in the United States or other Western or Western-aligned countries will require significant financial and temporal resources; therefore, Washington should not attempt to absorb Taiwanese workers before it builds the underlying infrastructure to support them. If labor flows too rapidly from Taiwan— or before the West constructs the infrastructure to efficiently leverage it— semiconductor output will decline and the global economy will suffer.

Government intervention would be needed to create alternative capacity in the absence of crisis in Taiwan, moving away from the existing supply chain structure that has developed through market factors. Such an intervention would need to be persistent to ensure success. The immediate need is to do the work to understand how much investment would be needed and how incentives should be structured to reshape the supply chain through intervention.

---

[4] Mark Tyson, "TSMC's Arizona Fab Hiring Woes Prompt Calls for Willing Taiwanese Migrants," Tom's Hardware, April 21, 2022; Yifan Yu, Cheng Ting-Fang, and Lauly Li, "From Somebody to Nobody: TSMC Faces Uphill Battle in U.S. Talent War," *Nikkei Asia*, May 7, 2022; Debby Wu, "TSMC Scores Subsidies and Picks Site for $12 Billion U.S. Plant," *Bloomberg*, June 8, 2020.

CHAPTER 6

Geopolitical Implications

Turbulence in semiconductor fabrication capacity is fundamentally a global economic challenge. But this turbulence also has potential military implications because of the military's dependence on economic strength and has implications for the geopolitical balance of power. The degree of interdependence between national economies is a new development, one that has created both mutual advantages and vulnerabilities. The TTX illustrated that because of integrated global supply chains, both economic and national security issues are inherent in any conflict with China. And the issues are likely opaque and difficult to predict in advance.

Because of the extreme concentration of global fabrication in Taiwan and the importance of semiconductors across the economy, economic vulnerability could provide the PRC with an asymmetric advantage. At a macroeconomic level, absorption of Taiwan would add about 3 percent to the PRC's gross domestic product. Taiwan is a wealthy society with a growing and vibrant economy, which would add to the PRC's overall national strength. Moreover, gaining control over TSMC's facilities would give the PRC a near monopoly over advanced semiconductor manufacturing, which confers global political and economic leverage. Although this does not confer complete dominance over the semiconductor supply chain since design capability stays largely in the West, it does give the PRC significant leverage over important parts of the process, given that the United States has relatively little manufacturing capability.

Peaceful Unification

In this scenario, China was able to acquire a significant portion of the semi-conductor global capacity without major cost to itself. The United States and its allies were faced with a near-term choice of accepting PRC dominance and continuing to work with Taiwanese companies now owned by China, or imposing sanctions and trying to cope with the loss of high-end production capability.

Industry and government players brought different perspectives. Although the industry players were aware that the rules for dealing with an authoritarian regime would be different—and potentially more restrictive—than engaging with democratic Taiwan, they felt that there was little option but to continue relationships with the companies now under CCP control. Indeed, industry specifically opposed ideas that would result in denying IP to these companies or inducing the movement of labor from Chinese-controlled Taiwan to the United States or its allies. Industry viewed U.S. government intervention that would restrict their access to Taiwanese supply as being more damaging than Chinese control of that supply.

The government players, conversely, felt that the risks associated with trading directly with CCP-dominated companies were enough to warrant strong and immediate action. However, if the United States attempts to impose countersanctions in nearly any sector, it is likely to run into resistance—within both the United States and other allied countries—to accepting economic pain to further a geopolitical end. Moreover, in the absence of prior investment to create alternative fabrication capacity, none of the actions recommended would have reduced vulnerability, at least not in the near term. The de facto result would be that the United States and its allies would have to accept the changed relationship and probably could not significantly reduce vulnerability for several years. Meanwhile, the PRC would stand to gain global influence as a result of possessing a near monopoly on at least the fabrication of the world's most sophisticated semiconductors.

If able to subordinate Taiwan through economic coercion, the PRC could challenge existing geopolitical boundaries throughout Asia. If the PRC could show its ability to withstand trade disruption at least better than the West, it could officially activate President Xi Jinping's plan to realize a Chinese-dominated, unipolar international system by mid-century.[1] Follow-on actions could include dispersion of PLA and People's Armed Forces Maritime Militia units throughout the South China Sea to test the territorial claims of Vietnam, the Philippines, and other neighboring states. Beijing could enhance its island reclamation projects in the South China Sea and defy Western calls to demilitarize the area. In all these cases, it will have demonstrated its ability to exert leverage, while resisting attempts to apply counter-leverage. This expanded influence could be expected to fundamentally change the global balance of power.

Contested Unification

In the contested scenario, the situation very rapidly became dire as access to Taiwan's semiconductor manufacturing quickly disappeared. The United States and its allies had no good nonmilitary options for dealing with the disruption, and conflict resolution came down to who would be better able to absorb the economic impact. The industry and government perspectives converged, with industry going as far as saying that government action would be needed to adjudicate prioritization of limited supply.

The contested scenario did not really leave an option for maintaining a status quo relationship with semiconductor manufacturers while beginning efforts to relocate production capability. The choices rapidly became

- cease supporting the Taiwanese effort to resist the coercive quarantine so that semiconductor production capacity could be restored through a Chinese-controlled Taiwan
- support the efforts and accept the loss of access to semiconductors and the loss of trade with the PRC, and thus face an economic depression
- consider military action to directly challenge and coerce the PRC.

Because the TTX did not include a military option, the choices in the contested scenario came down to capitulating and ceding significant geo-

---

[1] John Feng, "Xi Jinping Says China to Become Dominant World Power within 30 Years," *Newsweek*, July 1, 2021.

political influence to the PRC—and overriding the wishes of the Taiwanese people—or supporting Taiwan at the cost of a depression for most of the world and a years-long process of reshoring (absent prior planning).

The PRC's relative freedom of action and the inability of the United States and its allies to react to PRC actions—which was effectively the outcome of the TTX—would change the calculations of allies regarding the reliability and effectiveness of U.S. diplomatic and political guarantees. The United States will have been unable to respond to aggressive moves by the PRC, largely because it lacks the economic flexibility to respond.

The concentration of fab capacity also creates a potential risk for Taiwan, as opposed to a benefit. If the loss of Taiwanese semiconductor capacity means depression for the global economy, that might very well weaken U.S. and allied resolve. That would be particularly true if Taiwan does nothing to mitigate these outcomes or is perceived to be acting in ways that increase the threat to the global economy. Asking the United States and its allies to not just risk their militaries to defend Taiwan but to also live through a deep economic depression could be a major strain on alliance bonds.

A Call to Action

The outcomes and options in the TTX from both peaceful and contested unification scenarios were undesirable. The gameplay revealed that the United States is not ready to deal with any unification scenario, and no good short-term responses were identified. Without advanced planning to increase semiconductor production capacity outside of Taiwan, the United States and its allies will be in a very hard policy space should China move toward unification.

Military action is one policy option, but that is an undesirable path, even if it hadn't been removed from possible options for the TTX. Indeed, war with China would be a highly unwelcome outcome for the United States and its allies and for China. Absent military action, taking steps to support Taiwan could lead to broad and lengthy economic turbulence that would be politically unsustainable. Economic considerations among even the closest allies can lead to a reshaping of alliances if nation-state economies are at risk.

In the scenarios described in this report, advantage derives from being able to cope with disruptions to semiconductors produced in and exported from Taiwan. This could be a matter of having other sources of semiconductor fabrication, or it could be a matter of having an economic system better able to adjust to this particular shock. The countries that can most easily withstand disruptions to semiconductor capacity in Taiwan have an upper hand in strategic competition. If the United States and its allies have this advantage, it could be a powerful deterrent to Chinese action against Taiwan. The United States and its allies would be better able to mitigate the risk of global economic disruption and support Taiwan in resisting unification. If China has the advantage, it could act against Taiwan with reduced likelihood of interference from the United States and its allies to mitigate global economic risk.

In this TTX, the United States never gained an advantage and faced unfavorable outcomes in both scenarios. *This reality should be a call to action to assess options to increase fabrication capacity.* China is reportedly already taking such steps.[2]

---

[2] Karen M. Sutter, *China's New Semiconductor Policies: Issues for Congress*, Congressional Research Service, R46767, April 20, 2021.

CHAPTER 7

Recommendations

Throughout the first two decades of the 21st century, Americans have become accustomed to the benefits of rapid technological progress. The widespread availability of the internet and the development of the smartphone have helped to define the political, economic, and social fabric of the country. Thus, any major shock to the technology that underwrites the United States' modern way of life will fundamentally disrupt its society. Without a steady supply of high-end semiconductors, Americans will be denied access to cutting-edge healthcare equipment, suffer decreased work productivity, and lose social connectivity. Accordingly, Washington should do everything in its power to prevent supply chain disruptions or quickly resolve ones that do occur. A *do nothing* approach simply will not suffice for the United States; it will only breed social unrest and further destabilize a reeling country. **Thus, our recommendations for the executive branch of the U.S. government, the U.S. Congress, the governments of U.S. allies and partners, and industries with equities in the semiconductor supply chain center around increasing understanding, closing gaps, and changing incentives.**

1.   **Improve analysis and understanding of the semiconductor supply chain specifically and the overall level of supply chain interdependence in general**. The most obvious outcome from the TTX was discovering the need to improve the level of understanding of the vulnerabilities generated by the unique features of the semiconductor industry. The fact that high-end semiconductor production is uniquely concentrated in Taiwan, which adds to the vulnerability of these products, came as a surprise to many of the game participants. Two related but different lines of effort are needed.

a. First, the semiconductor supply chain is one in a list of supply chains whose poorly understood interdependencies were brought into focus by the COVID-19 pandemic. Companies may understand their individual supply chains, and even some executive agencies understand portions of the supply chains. Yet, during the TTX, few individuals participating had an appreciation for how quickly a crisis would hit or how severe its implications would be. Similar conditions could be present in multiple other sectors, but the research to establish these interdependencies has not been done.

b. Second, from a geopolitical perspective, planning scenarios involving conflict over Taiwan's autonomous status do not include the loss of Taiwanese semiconductor capacity as a likely consequence. This consequence deserves significant consideration.

2. **An immediate and concerted effort should be made to reduce the concentration of semiconductor production in Taiwan**. This condition is not only dangerous to the world's economic well-being, it also increases Taiwan's vulnerability. Reducing this concentration of semiconductor production will take several years, but this should be weighed against the PRC's own vulnerabilities. China's economy would be hurt by losing Taiwan's access and capacity. The management of vulnerability is, to a very large degree, a matter of timing. Several steps can be taken.

a. TSMC should be incentivized to move production out of Taiwan. This action does not imply moving all production, nor does it necessarily imply transfer of ownership. It means relocating production to places with less geopolitical significance than Taiwan. Reducing the risk of semiconductor production disruption because of Chinese aggression would increase the willingness of the United States and its allies to support Taiwan should aggression occur. This should be a powerful incentive for Taiwan.

b. Irrespective of TSMC actions, the U.S. and allied governments should take action to strengthen semiconductor production. Action does not imply top-down direction for investment, at least not in every case, but it does involve creating incentives for investment and creating opportunities for workforce training and liberalized immigration. It probably also involves management of IP-sharing with a clearer eye toward the security impacts of sharing designs, even those without an obvious defense tie. There might be designs that should be accessible only to producers inside the United States or preferred allies.

3. **Movement of facilities and equipment to the PRC should be specifically discouraged and heavily regulated**. If markets are incentivized to invest in the PRC and sell advanced equipment to Chinese companies, both are likely to occur, which furthers U.S. offshore dependence. Eliminating such incentives is likely to require coordination with allies and goes against the normal imperatives of a market economy. Incentives need to be structured in ways that industry will see as effective.

4. **Collaborative relationships with allied governments and industries are essential, even if these appear counter to the normal impulse to keep sectors separate**. The interdependencies created by supply chains are complicated and extensive, with individual and collective interests intertwining to a degree that means neither market nor normal government decisionmaking will be sufficient. This complexity requires extensive consultation, to the point that the relationships might necessarily be cozier than most democratic governments or private industries would prefer. The relationship between public and private sectors will require careful management, as will relationships with allies who have their own public-private challenges. But the TTX reinforced that neat separations between public and private interest are simply not possible in this context.

If a crisis arises, there might not be the luxury of time to seek additional information to inform decisionmaking. Thus, executive agencies need to engage with commercial industry to identify the strengths and vulnerabilities of the U.S. position within the global semiconductor supply chain. Specifically, new public-private working groups and communication channels can generate the panoramic *common operational picture* of the supply chain that business and government leaders currently lack. This informed view will guide investment decisions and prevent the executive branch from overreacting to minor shocks and underreacting to major ones. Such collaborations will be needed in other industry sectors too as supply chain interdependencies are better understood. Continued tabletop exercises of this kind, using different teams as ways of varying perspectives, might help in developing a framework for aligning private and public incentives. Furthermore, such repeated interaction between federal agencies and the country's leading microelectronics firms will build familiarity, or the *connective tissue* that is critical for expediting decisionmaking during a crisis.

Instead of acting unilaterally to potential disruptions to the semiconductor supply chain, Washington should engage its allies and partners to respond as a united, multinational bloc. Given the international nature of semiconductor production, potential U.S. actions—for example, halting the flow of IP to Taiwanese fabs— would generate cascading effects across the entire global economy. By engaging allies and partners in discussions about critical vulnerabilities and comparative advantages, the United States can better insulate itself from the economic fallout that would accompany any unforeseen disturbances to the semiconductor market. Whether the United States is the ideal country to lead this multinational bloc is in question. Because the United States has an adversarial relationship with the PRC, Beijing could see the United States leading this bloc as another sign of U.S. imperialism and a U.S. desire to control the international system. From a diplomatic messaging standpoint, the relegation of the United States to a secondary role could bolster the international legitimacy of the allied bloc.

Concluding Thoughts

RAND's *Assessing the Impacts of Interdependence* exercise was intended to add scholarly depth to the PRC-Taiwan dispute by investigating its challenges from broader economic and diplomatic perspectives. Although the exercise was not designed to identify a particular solution to the specific problems posed by either a peaceful or a contested unification, it has been contoured to spark meaningful interaction among the executive branch of government, the legislative branch, and private industry.

Overall, the *Assessing the Impacts of Interdependence* exercise triggered conversation and debate about the roles of the executive branch, the legislative branch, and private industry in a major supply chain disruption. Given the complexity of the crises presented and the facilitator-imposed communication challenges, we did not necessarily intend or expect to find comprehensive solutions to each problem set presented within the exercise. By fostering discussion, we sought to inspire the exercise participants to establish and improve communication about supply chain issues, in the pre-crisis phase. By bringing together key members of each interest group, we also built the first link in a potential chain of continued interactions. As these interactions—both interpersonal and interorganizational—mature over time, they will help form the connective tissue necessary for acting quickly and decisively if a crisis ultimately does occur.

It could not be clearer that serious vulnerabilities exist in the interrelationships resulting from the concentration of the semiconductor industry in East Asia, and, in particular, Taiwan. However, the system is considerably more complicated than any entity is likely to understand, even after an in-depth effort to do so. The potential for mistaken policy and counterproductive action is strong, *but the possibility of a mistake should not be taken as a signal to do nothing.* If effective action is not taken promptly, the ability to react to a geopolitical crisis might be reduced to poor choices indeed.

Abbreviations

AI	artificial intelligence
AMD	Advanced Micro Devices
CCP	Chinese Communist Party
CHIPS	Creating Helpful Incentives to Produce Semiconductors
COVID-19	coronavirus disease 2019
IP	intellectual property
OEM	original equipment manufacturer
OSAT	outsourced semiconductor assembly and test
PLA	People's Liberation Army
PRC	People's Republic of China
ROC	Republic of China (Taiwan)
SMIC	Semiconductor Manufacturing International Corporation
TSMC	Taiwan Semiconductor Manufacturing Company
TTX	tabletop exercise

References

Angell, Norman, *The Great Illusion; A Study of the Relation of Military Power in Nations to Their Economic and Social Advantage*, G.P. Putnam's Sons, 1910.

Biden, Joseph R., Jr., *Interim National Security Strategic Guidance*, White House, March 2021.

Bhutada, Govind, "The Top 10 Semiconductor Companies by Market Share," *Visual Capitalist*, December 14, 2021.

Boris, Teejay, "Qualcomm Snapdragon Chips Mostly Power $300+ Smartphones | MediaTek Tops Cheaper Devices," *Tech Times*, March 21, 2022.

Brock, David, and David Laws, "The Early History of Microcircuitry: An Overview," *IEEE Annals of the History of Computing*, Vol. 34, No. 1, January–March 2012, pp. 7–19.

Chang, Benjamin Angel, *Artificial Intelligence and the US-China Balance of Power*, dissertation, Massachusetts Institute of Technology, June 2021.

Chokkattu, Julian, "Qualcomm's New Smartwatch Chips Promise Big Battery Life Gains," *Wired*, July 19, 2022.

Companies Market Cap, "Largest Companies by Market Cap," webpage, undated. As of September 5, 2022: https://companiesmarketcap.com

Dean, Brian, "iPhone Users and Sales Stats for 2022," webpage, Backlinko, updated May 28, 2021. As of August 26, 2022: https://backlinko.com/iphone-users

Feng, John, "Xi Jinping Says China to Become Dominant World Power Within 30 Years," *Newsweek*, July 1, 2021.

Gibson, John E., William T. Scherer, William F. Gibson, and Michael C. Smith, *How to Do Systems Analysis: Primer and Casebook*, Wiley, August 1, 2016.

Google Finance, homepage, undated. As of August 26, 2022: https://www.google.com/finance

Guzeva, Alexandra, "'We Will Bury You': What Nikita Khrushchev Actually Meant," *Russia Beyond*, January 13, 2022.

Hof, Robert D., "Lessons from Sematech," MIT Technology Review, July 25, 2011.

International Data Corporation, "Worldwide Semiconductor Revenue to Grow 13.7%, but Supply Chain Remains Selectively Challenging Amidst Global Economic Volatility, According to IDC," webpage, June 8, 2022. As of August 26, 2022:
https://www.idc.com/getdoc.jsp?containerId=prAP49266822

Jie, Yang, "Fire at Giant Auto-Chip Plant Fuels Supply Concerns," *Wall Street Journal*, March 23, 2021.

Kilby, Jack A., "The Integrated Circuit's Early History," *Proceedings of the IEEE*, Vol. 88, No. 1, January 2000, pp. 109–111.

Martin, Bradley, Kristen Gunness, Paul DeLuca, and Melissa Shostak, *Implications of a Coercive Quarantine of Taiwan by the People's Republic of China*, RAND Corporation, RR-A1279-1, 2022. As of January 11, 2023:
https://www.rand.org/pubs/research_reports/RRA1279-1.html

Meredith, Sam, "Biden Says U.S. Willing to Use Force to Defend Taiwan—Prompting Backlash from China," CNBC, updated July 8, 2022.

Miller, Chris, *Chip War: The Fight for the World's Most Critical Technology*, Scribner, 2022.

National Institute of Standards and Technology, "CHIPS Act," webpage, updated April 5, 2022. As of August 26, 2022:
https://www.nist.gov/semiconductors/chips-act

Pettyjohn, Stacie, Becca Wasser, and Chris Dougherty, *Dangerous Straits: Wargaming a Future Conflict over Taiwan*, Center for a New American Security, June 2022.

Shepardson, David and Karen Pierog, "Foxconn Mostly Abandons $10 Billion Wisconsin Project Touted by Trump," Reuters, April 20, 2021.

Shlapak, David A., David T. Orletsky, Toy I. Reid, Murray Scot Tanner, and Barry Wilson, *A Question of Balance: Political Context and Military Aspects of the China-Taiwan Dispute*, RAND Corporation, MG-888-SRF, 2009. As of January 11, 2023:
https://www.rand.org/pubs/monographs/MG888.html

Sutter, Karen M., *China's New Semiconductor Policies: Issues for Congress*, Congressional Research Service, R46767, April 20, 2021.

Thomas, Jim, Iskander Rehman, and John Stillion, *Hard Roc 2.0: Taiwan and Deterrence Through Protraction*, Center for Strategic and Budgetary Assessments, December 2014.

Taiwan Semiconductor Manufacturing Company, "Logic Technology," webpage, undated. As of August 26, 2022:
https://www.tsmc.com/english/dedicatedFoundry/technology/logic

TSMC—*See* Taiwan Semiconductor Manufacturing Company.

Tyson, Mark, "TSMC's Arizona Fab Hiring Woes Prompt Calls for Willing Taiwanese Migrants," Tom's Hardware, April 21, 2022.

"U.S. Maintains 'Strategic Ambiguity' over Taiwan: Security Adviser," *Nikkei Asia*, July 23, 2022.

Varas, Antonio, Raj Varadarajan, Jimmy Goodrich, and Falan Yinug, *Strengthening the Global Semiconductor Supply Chain in an Uncertain Era*, Boston Consulting Group and Semiconductor Industry Association, April 2021.

Wayland, Michael, "Chip Shortage Expected to Cost Auto Industry $210 Billion in Revenue in 2021," CNBC, September 23, 2021.

Wu, Debby, "TSMC Scores Subsidies and Picks Site for $12 Billion U.S. Plant," *Bloomberg*, updated June 9, 2020.

Yeung, Henry Wai-Chung, "Explaining Geographic Shifts of Chip Making Toward East Asia and Market Dynamics in Semiconductor Global Production Networks," *Economic Geography*, Vol. 98, No. 3, 2022, pp. 272–298.

Yu, Yifan, Cheng Ting-Fang, and Lauly Li, "From Somebody to Nobody: TSMC Faces Uphill Battle in U.S. Talent War," *Nikkei Asia*, May 7, 2022.